V1 The Flying Bomb

Joachim Engelmann

SCHIFFER MILITARY HISTORY
Atglen, PA

Translated from the German by Edward Force.
Copyright © 1992 by Schiffer Publishing, Ltd.

ISBN: 978-0-88470-408-5
Printed in China.

This title was originally published under the title,
V 1 Die Fliegende Bombe.
by Podzun-Pallas Verlag, Friedberg.

We are always looking for people to write books on new and related subjects.
If you have an idea for a book, please contact us at the address below.

Schiffer Books are available at special discounts
for bulk purchases for sales promotions or
premiums. Special editions, including personalized
covers, corporate imprints, and excerpts can be
created in large quantities for special needs.
For more information contact the publisher:

Published by Schiffer Publishing Ltd.
4880 Lower Valley Road
Atglen, PA 19310
Phone: (610) 593-1777;
Fax: (610) 593-2002
E-mail: Info@schifferbooks.com

For the largest selection of fine reference books on
this and related subjects, please visit our web site at:
www.schifferbooks.com

We are always looking for people to write
books on new and related subjects. If you have an idea
for a book please contact us at the above address.

This book may be purchased from the publisher.
Include $5.00 for shipping. Please try your
bookstore first. You may write for a free catalog.

In Europe, Schiffer books are distributed by
Bushwood Books
6 Marksbury Ave.
Kew Gardens
Surrey TW9 4JF England
Phone: 44 (0) 20 8392-8585;
Fax: 44 (0) 20 8392-9876
E-mail: info@bushwoodbooks.co.uk
Website: www.bushwoodbooks.co.uk

Gerhard Fieseler, born 1896, first world champion in precision
flying, defeating Ernst Udet, and since 1930 Director of the
Fieseler Aircraft Construction Corporation, Kassel, creator of
the first low-flying Fi 156 "Storch" (Stork) airplane and
organizer of the V1 (Fi 103) project for the "flying bomb."

Design and Development of the V1

In August of 1939, Dr. Ernst Steinhoff, division director at Peenemünde-Ost, submitted to the Reich Air Traffic Ministry a memorandum on "Flights at Enemy Targets with Unmanned Aircraft", with the suggestion that unpiloted bombers be directed to their targets by utilizing enemy radio transmitters with special radio navigational processes. There had already been initial tests of this in 1930, and in 1937 there had been flight testing of flying objects with autopilots made to designs by the "German Experimental Office for Air Travel." The Ministry took no action. The offer, submitted again in 1941, was strictly declined, because on September 11, 1941 Hitler had banned long-term weapon development.

When the development of the V2 seemed to be getting nowhere in 1942 and Germany's air raids on England became more and more costly — and "retaliation raids" for British attacks on German cities were to be made for the purpose of making Britain "more ready for peace", the idea of long-range attacks with new weapons and large quantities of explosives popped up again. It culminated in a small unmanned aircraft intended to be lost, inexpensive, remote-controlled, with 800 kilograms of explosive and a 250-kilometer range, for use against a surface target of 20 by 10 kilometers, thus a ground-to-ground missile. A special kind of propulsion had to be employed, and it had to take off from a launching area or carrier airplane. Its speed was limited, its flying altitude was between 300 and 2500 meters. The project was directed completely at London and southern England. The remaining problems were disturbances to the originally planned remote control and the precision factor in attacking a target with large length and breadth dispersion. But it would save pilots.

The propulsive power was to be the pulsating athodyd tube secretly and successfully tested in a Dornier Do 17 Z by Eugen Sänger on March 7, 1942 as a "stovepipe", developed for the Army Weapons Office by Paul Schmidt since 1937, and first used in practice for the Argus Motor Works on a Gotha Go 145 by G. Diedrich on April 28, 1841. This propulsion system was a pioneering invention and the simplest powerplant for aircraft, without moving parts, but it required oxygen from the air to run and was thus limited to the earth's atmosphere and thus unusable in space. The compressed air from outside entered through openings in the surface and, after having fuel sprayed into it, was ignited with closed vents in such a way that the exhaust gases streaming out the end of the tube created a strong thrust, the vacuum of which immediately sucked new fresh air into the front from outside. This process pulsed 50 to 250 times per second. Thus it was designated "propulsion motor." The thrust was therefore dependent on the flying speed. The higher the speed, the better the powerplant worked; at rest there was zero thrust. Therefore the effective speed necessary from the start had to be created by massive take-off

General der Flakartillerie Walther von Axthelm, as Inspekteur der Flakartillerie, promoted the development of the V1 decisively. He had previously been Commander of the "General Göring" Flak Regiment, the 1st Flak Brigade, and the I. Flak Corps.

assistance via catapult, rockets or carrier planes. The greater the oxygen density, the better the powerplant functioned, which called for low flight altitudes. The powerplant ran for 30 minutes. The disadvantages were the easily located noise of combustion, the visible explosion gases and the speed, which was less that of fighters and other warplanes.

Since the stability of the craft was only achieved at 300 meters per second, it had to be catapulted off a ramp at least 35 meters long with enough force so that at the end of the ramp, the necessary speed was sure to have been attained. With a take-off weight of 2.2 tons, that took a bit of power! A counter propeller on the nose measured the distance covered and released the setting of the elevators, the fuel stop and the wings, which were jettisoned. Under the weight of the explosive payload, the rocket tipped forward and plunged to the ground as at the end of a trajectory parabola. The control was only suitable when it was very precise and robust. Every slightest operational error ar failure of the three gyrocompasses led to failure at the take-off or shortly thereafter, very important in the case of this weapon's new nature and not yet attained readiness for action. The design by Dr. Robert Lusser of the Fieseler Works in March of 1942 received, on June 10, 1942, not only immediate approval after only five days but "highest priority" for development and production, thanks to the involvement of Generalfeldmarschall Erhard Milch, so that the Luftwaffe would now have its own long-range weapon developed from aircraft construction. The Fieseler Works took over the whole direction and production of the outer missile, Askania provided the steering and control system, Argus provided the

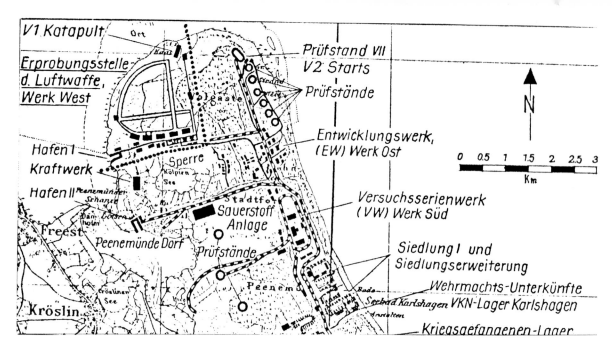

On the northwest rim of the Peenemünde Point lay the "Test Position of the Luftwaffe", West Works, very near the Army Test Center and easy to protect. Here the Fi 103 was developed in a very short time and also tested (After many bomb and weapons developments, this was the first large-scale project of the Luftwaffe at Peenemünde).

powerplant, Peenemünde carried out the testing, and the Volkswagen Works produced the cells from their production lines, first in Wolfsburg and later in their central plant in the southern Harz Mountains. In December of the year the models were tested by being released from a Focke-Wulf Fw 200 C and on December 24, 1942 the first ground take-off took place, a tremendous achievement for the planners, engineers and workers!

Although after the first take-off in early January of 1943, the first test launches in the presence of Hitler, Göring and Himmler were not very convincing, and those shown to Speer, Milch and Dönitz on May 26, 1943 failed, the mass production of the missile was decided on, and the Instruction and Testing

Command was set up at Zinnowitz-Zempin under Oberstleutnant Max Wachtel, whose personality impelled the practical work to move ahead smoothly. As with the A4, the flight path ran along the Pomeranian coast to the level of Rügenwalde, controlled by measuring stations. On June 27, 1943 a long-distance launch traveled 234 kilometers, soon followed by one of 242 kilometers; then came launches to an altitude of 2300 meters and a speed of 625 kph. On June 28, 1943, at Speer's suggestion, Hitler ordered the construction of four massive, laboriously built concrete launching bunkers as part of the "Atlantic Wall." Soon 252 sites were announced, and a great many of them were already under construction. Their area extended from south

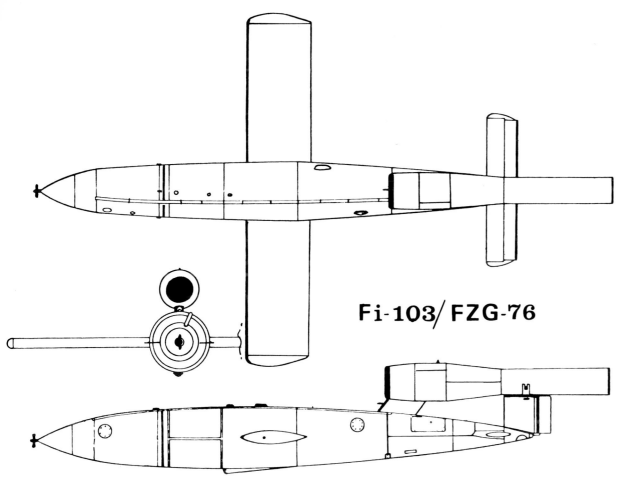

Fi-103/FZG-76

The top, side and end views of the Fi 103/FZG 76 show its uncomplicated, simple form, completely sufficient for the "launched small aircraft" insofar as it fulfilled the aerodynamic requirements. As opposed to normal aircraft, the rudder is poorly developed, all organs serve propulsion and launching. It was surprising that it could carry more than one-third of its gross weight as a payload.

of Dunkerque between Amiens and Abbeville, Rouen and Le Havre through Caen to the peninsula of Corentin. River valleys, hilly land, forests and scrub areas offered good potential for installation and camouflage, although the firing positions, each with twelve to thirteen structures, were not hidden from enemy air reconnaissance, nor from

espionage, especially on account of the long storage bunkers with their curved entries. Also typical were the launching ramps, originally made of concrete. On September 23, 58 of 64 main launching positions were ready, and 32 changeable positions were underway. On the other hand, as of December 5, 1943 the Royal Air Force began systematic

attacks on all launching ramps, production plants and developmental facilities.

Preparation for firing consisted of three separate areas:
1. Assembly and preparation of the flying bomb,
2. Storage and readiness maintenance,
3. Carrying out the launching.

Next to the servicing and unloading platform at the entrance were two or three workshop buildings, partly underground, for various purposes. Then the V1 went into the assembly hall, where it was assembled and equipped with a powerplant and controls. After being fueled and filled with compressed air, it was taken to one of the three storage bunkers, each of which housed ten missiles. In addition, there was also a fuel bunker of 24 cubic meters, a technical bunker for water pumps, heat, electricity and generators, as well as a water basin for cooling water and fire extinguishing. Before the launch, the V1 entered the anti-magnetic aiming house, where the gyrocompass and flight time were set and the last checks made. Attaching the wings and inserting the fuse made it ready to launch. From the launching platform it went to the catapult, which was 48 meters long and raised five meters at the end. To the left rear of it, the shot leader activated the launch electrically with a portable switch from the concrete command post. As of February 1944, enemy air attacks compelled the fortified positions to be vacated, especially the storage bunkers, and the transition to constructed steel catapults, which made most of the buildings unnecessary and changes of position easier.

After June 3, 1944 its first performance went off briefly. The opening of fire demanded by the OKW on June 13 was technically premature and the necessary preparations were hastened; the difficulties were overcome successfully in a few days. The men of the regiment were overburdened in many ways. Accidental losses sometimes outnumbered losses to enemy action. At first the regiment consisted of four units, each with four launching batteries and two technical batteries, making 64 launchers in all. Every battery had about 215 men, the unit 1400, the regiment about 5700. The launching crews included four Wachtmeister, five Unteroffiziere and forty men, divided into assembly, launching, loading and transport groups and radiomen. As of July 21, Flak Regiment 255(W) joined them to secure the transportation of the missiles, a difficult job in view of enemy air superiority.

The best time for launching was at night or under heavy cloud cover, so as to prevent premature spotting of the missiles. The number of failures and strays was a problem that could be lowered only gradually. The explosive effect of the bomb was increased by new types of explosive; it was higher than the British censorship and propaganda could admit, though not as decisive as had been hoped. The efforts of the British government are clear indications of this. In addition, considerable efforts were devoted to defenses against the bombs. When the fortunes of war in general decrease, one weapon, no matter how new, cannot be more successful than the sum total.

The Allied breakthrough to the Seine on August 20, 1944 destroyed the security of the launching positions along the Atlantic coast. The lengthy change of positions was covered

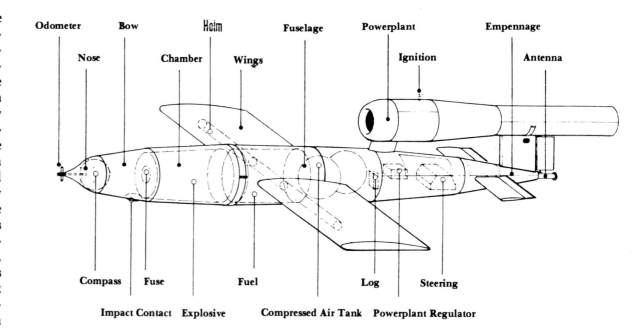

This drawing shows the location of the technical devices and division of space in the missile, with the center of gravity before the wings, which was equalized during flight.

by the use of V2 rockets and launchings from He 111 planes by KG 53. By then, 8892 V1's had been launched. Of them, 1500 V1's had been shot out of the air over England. But the biggest problem was the insufficient production, which only reached the level of 3000 in September of 1944, instead of the prescribed number of 9000, and as of November production decreased again. It was affected by air raids, Hitler's interference and production delays. After the 10,000th V1 was launched on November 19, 1944, the regiment had to vacate Roubaix. London was now out of range. As of December 16, 1944 the regiment fired from Holland on Antwerp, Brussels and

Liege. After air action was halted at the end of January 1945, new launches against London and southern England were commenced on February 20, 1945, now directed by the SS and using 370-kilometer "range cells." When the war ended, the 5th Flak Division (W) was located between Münster and Hamburg. According to British calculations, the ratio of German effort to enemy damage was 1 to 4, although the V1 was not at all fully developed technically.

What could have been accomplished was shown by developments between 1945 and 1976: Cruise missiles.

Testing

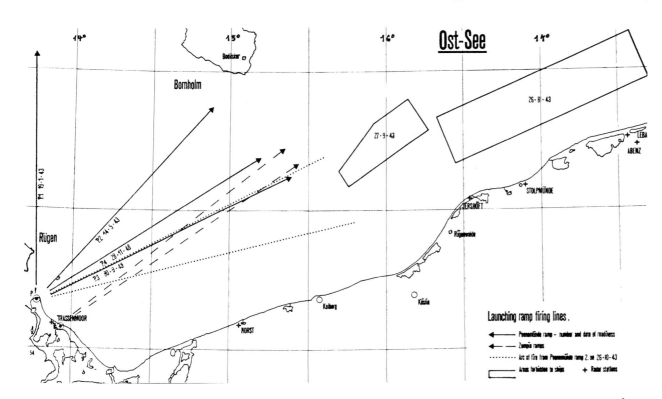

Left: The firing ranges for testing, used from January 19 to September 30, 1943, turned further and further to the northeast along the Pomeranian coast, the longer the distances and the more precisely they were measured.

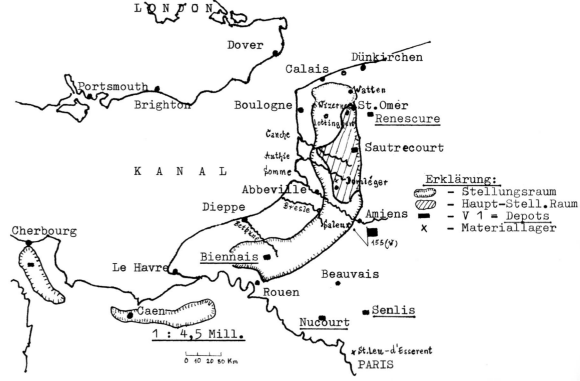

Right: After long-term reconnaissance, the construction of launching sites began prematurely, but all too expensively and vulnerably, between Calais and Cherbourg, with flight paths concentrating on Greater London. The weapon was meant to cause political terror, though military targets were more important.

Position Construction on the Channel Coast, 1943-44

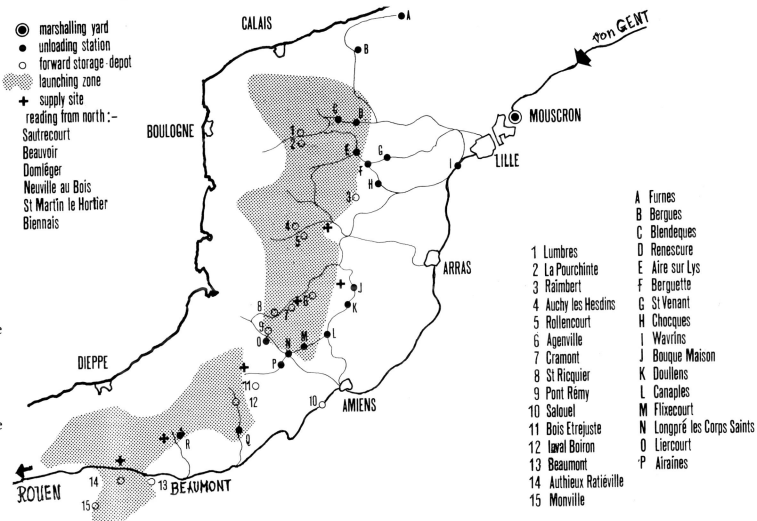

marshalling yard
unloading station
forward storage depot
launching zone
+ supply site
reading from north :-
Sautrecourt
Beauvoir
Domléger
Neuville au Bois
St Martin le Hortier
Biennais

Corresponding, the supply system ran from the Ghent-Rouen railroad line to the Channel coast into the deep river valleys that led to the coast. Without the very vulnerable but well-organized supply system, the regiment would not have been ready for action. Enough difficulties occurred as it was.

CALAIS
BOULOGNE
DIEPPE
ROUEN
von GENT
MOUSCRON
LILLE
ARRAS
AMIENS
BEAUMONT

A Furnes
B Bergues
C Blendeques
D Renescure
E Aire sur Lys
F Berguette
G St Venant
H Chocques
I Wavrins
J Bouque Maison
K Doullens
L Canaples
M Flixecourt
N Longpré les Corps Saints
O Liercourt
P Airaines

1 Lumbres
2 La Pourchinte
3 Raimbert
4 Auchy les Hesdins
5 Rollencourt
6 Agenville
7 Cramont
8 St Ricquier
9 Pont Rémy
10 Salouel
11 Bois Etrejuste
12 Laval Boiron
13 Beaumont
14 Authieux Ratiéville
15 Monville

Nachschub-System für die V 1 - Stellungen
Eisenbahn - Ausladestationen - vorgesch. Depots -
Nachschublager

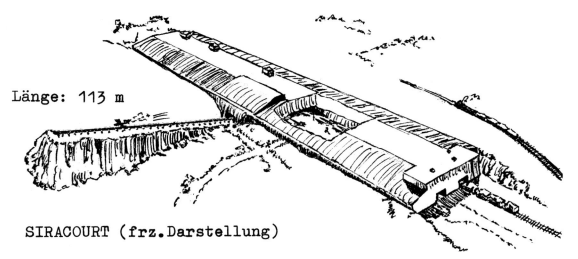

Länge: 113 m

SIRACOURT (frz.Darstellung)

Above: Siracourt is an example of the large buildings with bomb-safe quarters that were originally erected. Launching and underground railroad supply lines: barracks, assembly, storage and launching under one roof, but bound to the location with all its disadvantages. The other large-scale buildings were similar in concept.

Above: The semi-official "Organization Todt" handled the employment of civilian construction firms with natives of the country as workers; they were treated, like soldiers, as "those required to serve."

Ventilations-Blocks

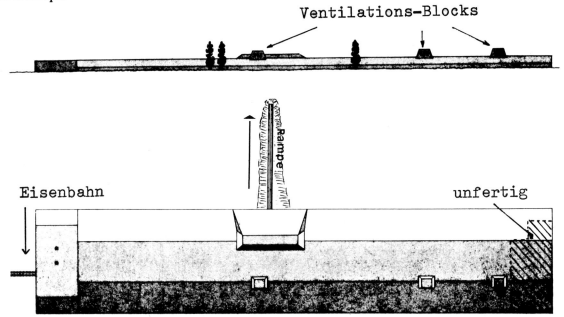

Eisenbahn

Rampe

unfertig

Left: This drawing shows (seen from above) the present-day state of the Siracourt facility, which was never used in practice, and (seen from the side), the ventilation shafts, the only components visible above ground, a senseless expenditure in unfinished space, with a reinforced concrete launching ramp.

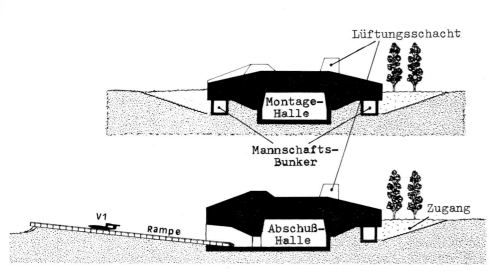

Above: Assembly hall with personnel quarters and launching hall were offset to the side and separated from each other for reasons of safety and space. These structures, results of "fortress thinking", were unsuitable for combat use.

Upper right: These are not battle stations, but the ventilation shafts at the Siracourt facility. Everything else has disappeared underground, a gigantic hall 113 meters long with several floors — almost like a railroad station.

Right: The use of well-camouflaged, spread-out field positions with considerably simplified concrete structures, spread out in larger numbers, as shown here in a typical design. They also served as switching points.

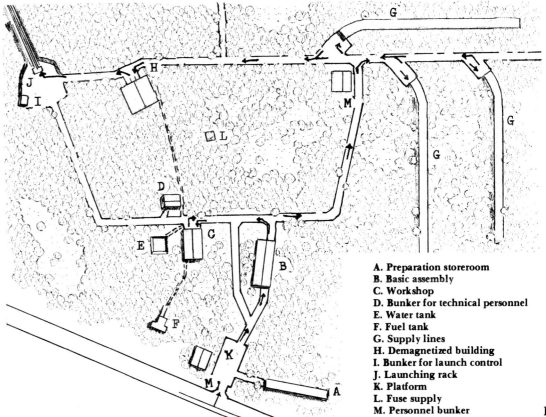

A. Preparation storeroom
B. Basic assembly
C. Workshop
D. Bunker for technical personnel
E. Water tank
F. Fuel tank
G. Supply lines
H. Demagnetized building
I. Bunker for launch control
J. Launching rack
K. Platform
L. Fuse supply
M. Personnel bunker

11

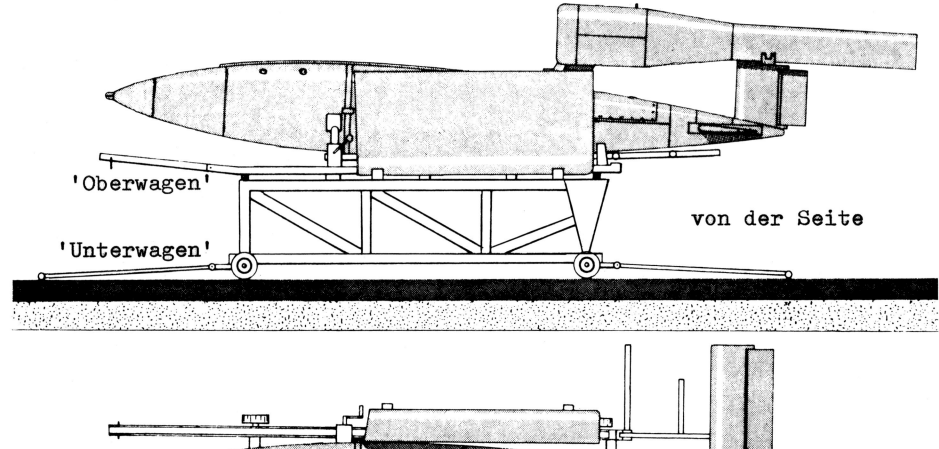

'Oberwagen'

'Unterwagen'

von der Seite

V1 auf Transport-Karren
von oben

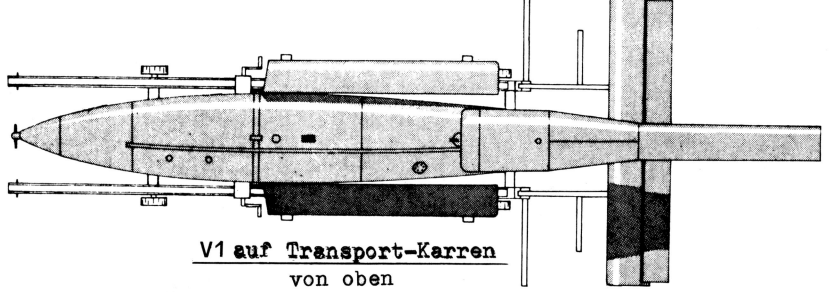

After delivery, the rockets were moved inside the launching position only on transport carriages, with detached wings and unfueled, after they had been unloaded and basically assembled in the workroom.

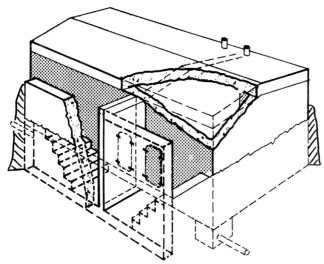

TREIBSTOFF – Bunker für den Dampferzeuger mit getrenntem hochprozentigem Wasserstoff-Superoxyd und Kalzium-Perman=ganat als Katalysator

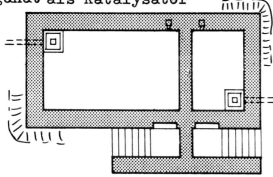

Above: The fuel tank was isolated — made of massive reinforced concrete — and by far the most solidly constructed building at the site: it measured 24 cubic meters and pumped the fuel to the workshop underground.

Right: This is how they looked from outside, half underground, here still with a level entrance that was changed later. Every site had several such bunkers.

Querschnitt durch Lager-Bunker

Above: Ten V1 with detached wings were housed safely in these storage bunkers.

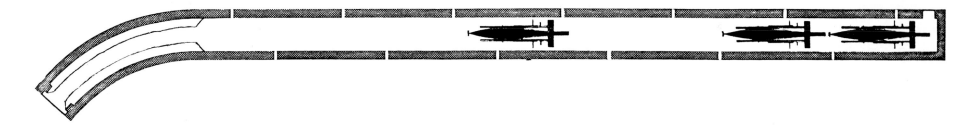

The entrances, later curved for shrapnel protection, betrayed their nature to enemy air reconnaissance by being so conspicuous, and were identified easily.

Lower left: The anti-magnetic "aiming house", central point of the site, was where the V1 was completed and equippedwith gyrocompass and flight-time programming for targeting before it reached the launching platform.

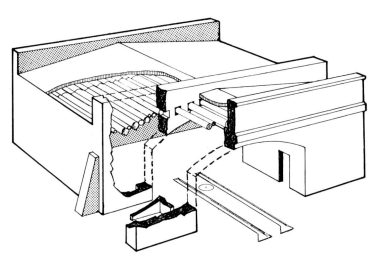

Above: Cutaway drawing of an "aiming house" (anti-magnetic) with parallel structures and arches, for compass installation.

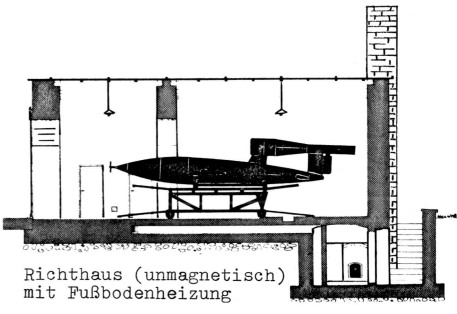

Richthaus (unmagnetisch)
mit Fußbodenheizung

Above: The interior of the "aiming house" shows the typical rounded arches and the rails parallel to the launching direction of the missile.

Upper right: Floor heating provided dry air and even temperatures independent of the heat outside. Preparations concluded with the attachment of the wings and installation of the fuse.

Right: The "aiming house" can be seen clearly at right; to the left of it is the launching bunker, in front of it the approach to the catapult system.

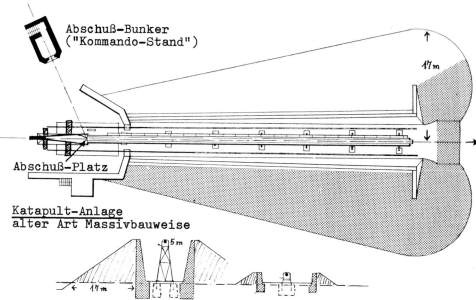

Abschuß-Bunker
("Kommando-Stand")

17 m

Abschuß-Platz

Katapult-Anlage
alter Art Massivbauweise

5 m

17 m

This is how a sidewall of the massive concrete launching ramp looked: 48 meters long, with 6-degree inclination and 5 meters high at the end, to attain the necessary take-off speed, a gigantic structure.

The complete take-off facility looked gigantic from above the from the side (as shown here), with a surface area of 48 x 34 meters, and there were four of these in each site, sixteen in each unit!

Querschnitt des Abschuß-Bunkers mit Kommando-Pult

Left: The small launching bunker, half-underground and completely of reinforced concrete, was situated in a dead angle to the left rear of the launcher, with electric ignition and alarm system.

Simplified Firing Positions

Betoniertes Straßenkreuz als vereinfachte V 1 - Stellung

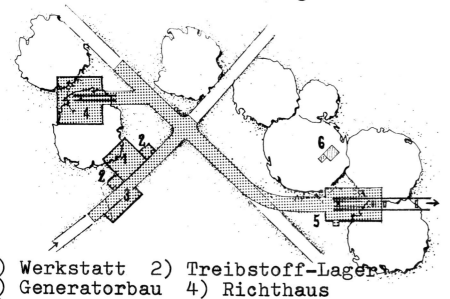

1) Werkstatt 2) Treibstoff-Lager
3) Generatorbau 4) Richthaus
5) Abschuß-Platz 6) Kommando-Stand

Above: On account of continual enemy bomb attacks, simplified site construction and greater mobility had to be adopted.

Right: Simplified construction methods without covering were also used for quicker departure and better camouflage via underground switching, culminating in the filigree and slit construction pattern.

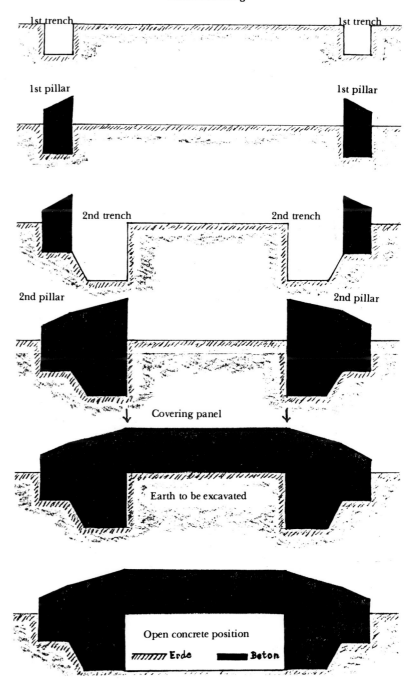

FIELD BUNKERING OF V1 LAUNCHING SITES
without covering

1st trench — 1st trench

1st pillar — 1st pillar

2nd trench — 2nd trench

2nd pillar — 2nd pillar

Covering panel

Earth to be excavated

Open concrete position

/////// Erde   ▬▬ Beton

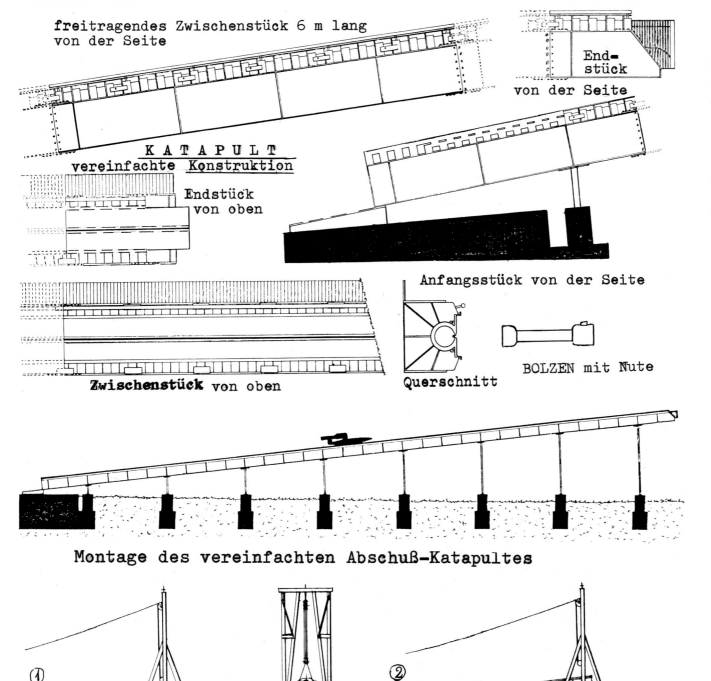

freitragendes Zwischenstück 6 m lang
von der Seite

End-
stück

von der Seite

KATAPULT
vereinfachte Konstruktion

Endstück
von oben

Anfangsstück von der Seite

Querschnitt

BOLZEN mit Nute

Zwischenstück von oben

Montage des vereinfachten Abschuß-Katapultes

①　②

The simplified catapult without sidewalls consisted of a standardized steel framework, of which only the beginning and end were made of concrete. It was portable and could be dismantled. The bolt with nut protruded into the rocket's path to the front.

Launch of a V1 from the simplified catapult.

With cranes and lifting booms the new catapult could be put together for field use, assembled in pieces and set up.

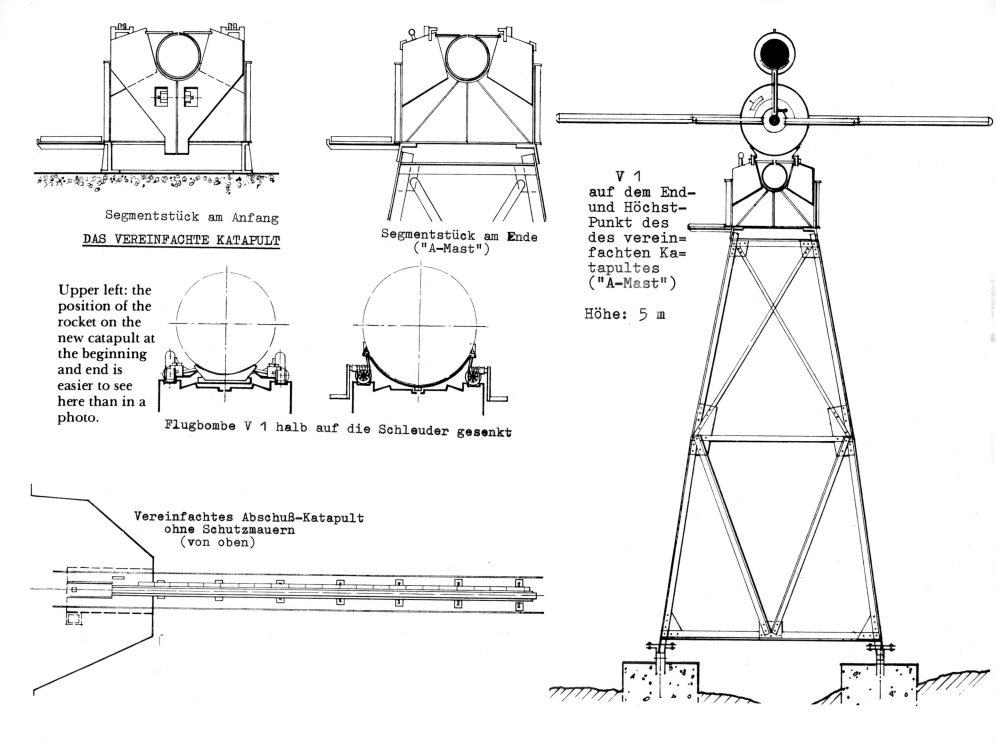

Segmentstück am Anfang

DAS VEREINFACHTE KATAPULT

Segmentstück am Ende
("A-Mast")

Upper left: the position of the rocket on the new catapult at the beginning and end is easier to see here than in a photo.

Flugbombe V 1 halb auf die Schleuder gesenkt

Vereinfachtes Abschuß-Katapult
ohne Schutzmauern
(von oben)

V 1
auf dem End-
und Höchst-
Punkt des
des verein=
fachten Ka=
tapultes
("A-Mast")

Höhe: 5 m

Left: Combat areas of the units of Flak Regiment (W) 155 with command posts in June 1944 (from original map in Federal Archives), from Ostend to Cherbourg, with their lines of fire.

During firing pauses, tractors brought new missiles up from the storage bunker to replace those that were launched.

Operations

From the "aiming house" the ready-to-launch V1 is moved manually to the launching place on a carriage, as seen here on August 8, 1944. It was a great physical strain for the soldiers.

Upper left and above: The approach paths were often protected by concrete coverings. It was not easy to move the rocket's weight of more than two tons on these small wheels. (1 x BA)

Left: The crane lifted the V1 off the transport carriage and set it on the catapult, while the controls for the horizontal flying altitude were set at the tail.

For bursts or sustained fire, additional rockets were kept ready before the launching place. (BA)

Above: Now the rocket sat on the starting point of the catapult for attachment.

Left: While additional rockets, seen here on single-axle trailers, stood ready in the background, fourteen men pushed the V1 forward on the catapult so it could be bolted.

Right page: After being bolted fast, the control device and the range programming were checked one last time to avoid failures, premature landings or circular courses.

Above: The Heinkel steam generator ("baby carriage") made the Walther mixture react at the rear end of the tube and drove the block through the tube like a shell with a great cloud of steam. It flew out with the V1 and had to be recovered for reuse. The Borsig catapult by Friedrich Clar, with a smokeless solid-fuel rocket, had been rejected by the Air Ministry.

Left page: The damaged catapult in an abandoned firing position still provides a full impression of the weapon, which was then top secret, and its launching technology.

Above: The launching ramp of the catapult at Belloy-sur-Somme in this picture shows the groove for the bolt head between the guiding rails.

Below: A very rare photo: A British reconnaissance aircraft photographed the rocket on the catapult before launching, although launching was usually done in bad weather and under low clouds.

In Flight

Left: Ascending flight of the V1 after launch. (DM)

The V1 in horizontal flight toward a distant target, its powerplant howling fiercely.

Right: The flight paths recorded by British intelligence on August 30, 1944 — two days before the regiment's great position change after the Allied breakthrough across the Seine — show firing positions and targets. The four units had launched 8617 V1 rockets from up to 64 launchers since June 1, 1944. The strength of the regiment was at most 6800 men.

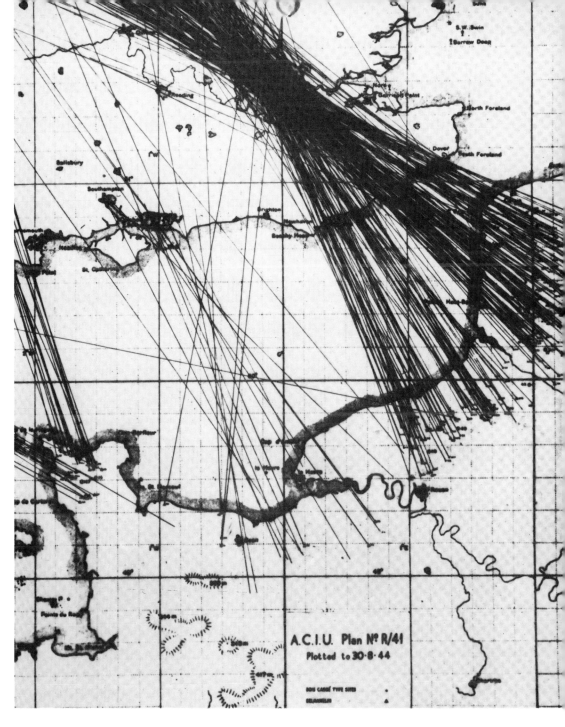

A.C.I.U. Plan Nº R/41
Plotted to 30·8·44

Manned V1 ("Reichenberg")

FZG 76 "Reichenberg" (Type 2) with sprung skid and landing flaps as a single-seat practice craft, in flight without a payload.

Above: The craft flown by pilots since testing in August of 1943 were developed further as the "Reichenberg", of which there were three versions. They were intended for later "suicide flights" by some 200 craft. The first to fly one was Willy Fiedler, Technical Director of this secret project, then Hanna Reitsch, flying from the Rostrup Air Base near Zwischenahn. (BA)

Right: Type 3 (without wings) sits on a transport carriage with no way to land; it was a ground-to-ground or air-to-ground missile, launched from a Heinkel He 111.

Americans gaze in wonder at a Type 3 "Reichenberg" in 1945, describing it as a "last-ditch weapon" or "desperation weapon" that, despite the readiness of 15,000 volunteers, Hitler did not want to use, so that at the end of 1944 the secret supply was blown up. (BA)

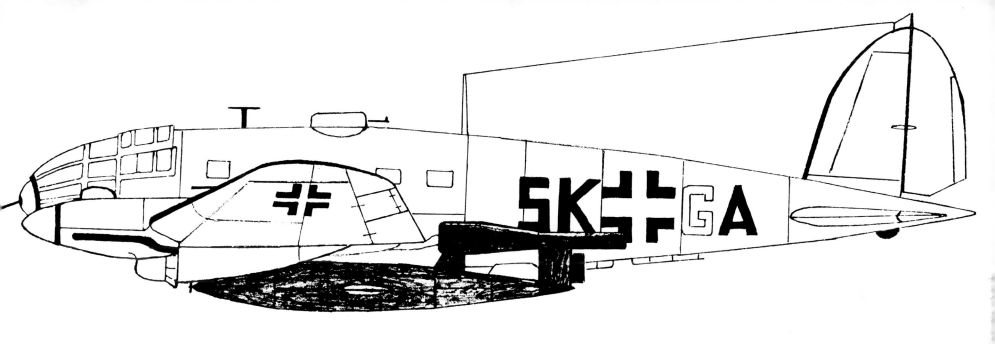

Launch from the Air

Above: Drawing of the He 111 H 22 with a V1 hanging under its left wing — the difference in size shows clearly that the flying missile was just a weapon, scarcely an aircraft, compared to the carrier plane.

Right: Since September 1944 III./Kampfgeschwader (KG) 3, since November of that year the entire KG 53 at Varrelbusch, Bad Zwischenahn and Leck, launched the V1 against England from the air till the war ended. Here the rocket is seen hanging under the left wing of a Heinkel He 111.

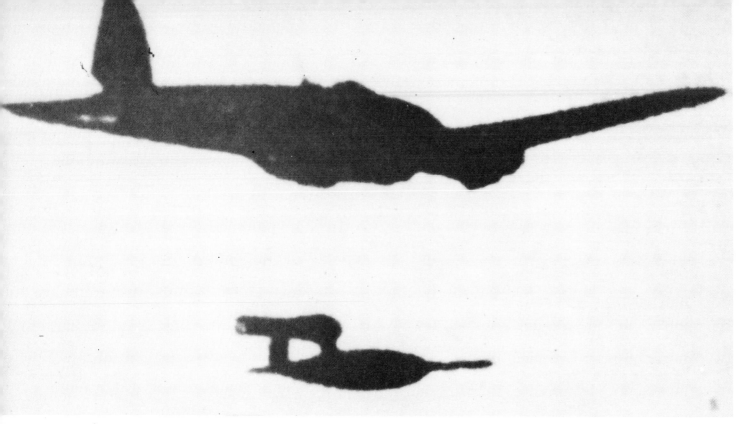

This rare photo shows the launch of an FZG 76/V1 from an He 111 at the moment when the necessary launching speed of the missile has been attained and it can fly on under its own power.

The Arado Ar 234c2 jet bomber carried the V1 piggyback — the raised launching position is shown — but also launched "glider" V1 versions without motors or elevators, but with fixed controls, as air-to-ground missiles.

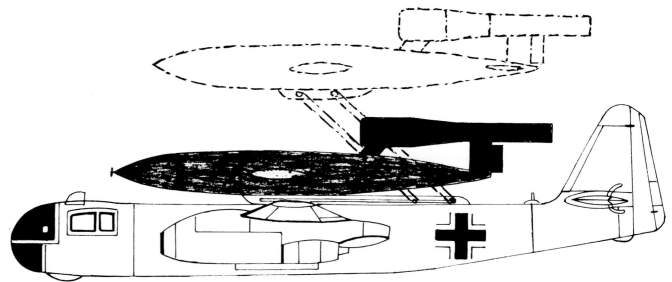

ARADO Ar.234c2

V1 Defense

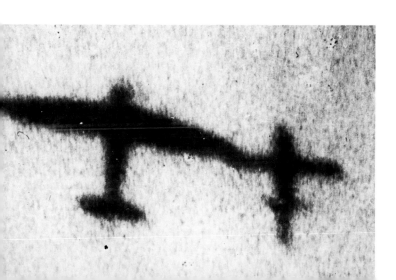

Above: This is how a pursuing British fighter saw the V1 at the moment of attack.

British fighters (right) shot down the V1 from a superior altitude. Daring pilots (below) tipped it out of its course and equilibrium. (1 x BA)

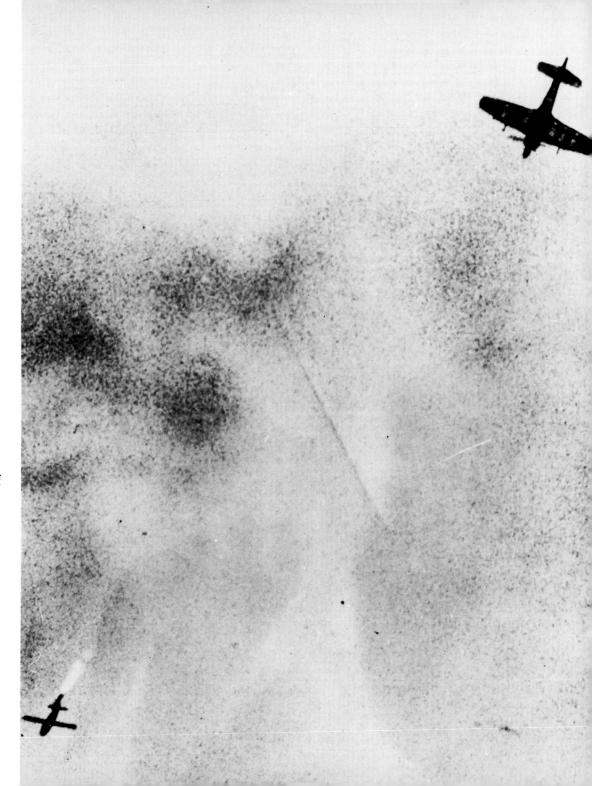

(Seltsam Keit) ODDENTIFICATION

"Doodle Bug!"
(Kritzel-Gespenst)

Above: This caricature shows that the British did not regard the V1 as a decisive weapon.

Right: The acoustic fuse of the British anti-aircraft shell said to the V1 motors at a range of 300 meters: "Death to the flying bombs!"

Left page: After reaching target range, the elevators angled off, stopped the fuel and pushed the wings away — and the V1 fell. (1 x BA)

THE ACOUSTIC SHELL
(Schallempfindlicher Granat-Zünder)

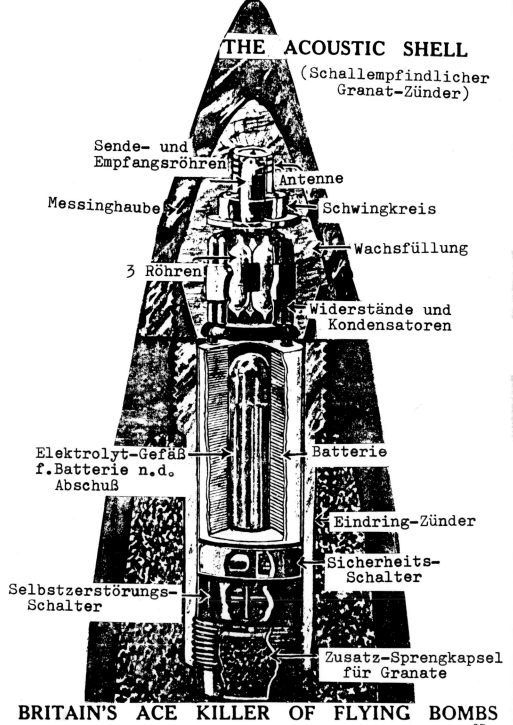

Sende- und Empfangsröhren

Antenne

Messinghaube

Schwingkreis

Wachsfüllung

3 Röhren

Widerstände und Kondensatoren

Elektrolyt-Gefäß f. Batterie n.d. Abschuß

Batterie

Eindring-Zünder

Sicherheits-Schalter

Selbstzerstörungs-Schalter

Zusatz-Sprengkapsel für Granate

BRITAIN'S ACE KILLER OF FLYING BOMBS

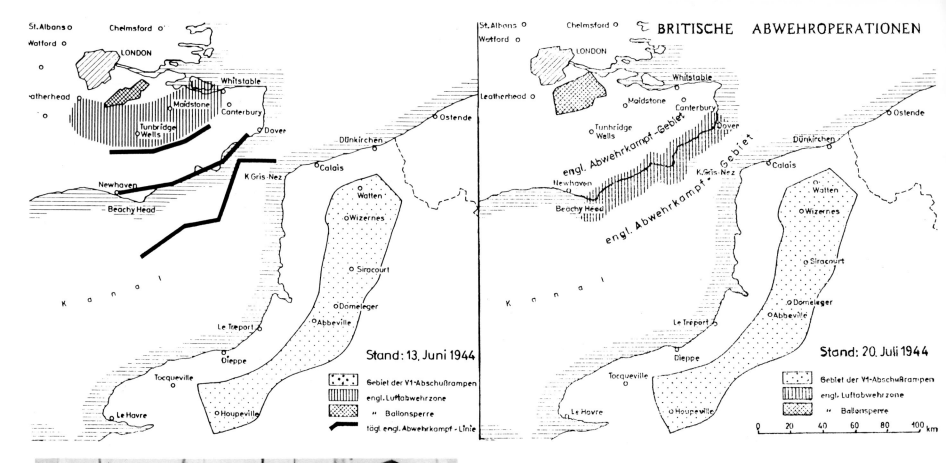

BRITISCHE ABWEHROPERATIONEN

Stand: 13. Juni 1944

- ⣿ Gebiet der V1-Abschußrampen
- ▥ engl. Luftabwehrzone
- ▦ „ Ballonsperre
- ▬ tägl. engl. Abwehrkampf-Linie

Stand: 20. Juli 1944

- ⣿ Gebiet der V1-Abschußrampen
- ▥ engl. Luftabwehrzone
- ▦ „ Ballonsperre

engl. Abwehrkampf-Gebiet

0 20 40 60 80 100 km

BALLOON ARMAMENT

Above: Within six weeks the British were able to reorganize their air defenses by concentrating and delaying their balloon barrages and advancing their defense to gain time to fight the V1.

Left: Displaying a dud in Piccadilly dispersed the mysterious air of the V1 among civilians.

Right: As the map shows, the strays, duds and premature falls reached about 35%, a figure later reduced to about 8%. Mistakes made by the overworked crews or material failures were the usual reasons. But there were also accidents to the weapons. The regiment lost a total of 189 dead, 321 wounded, 71 missing — more than 10%.

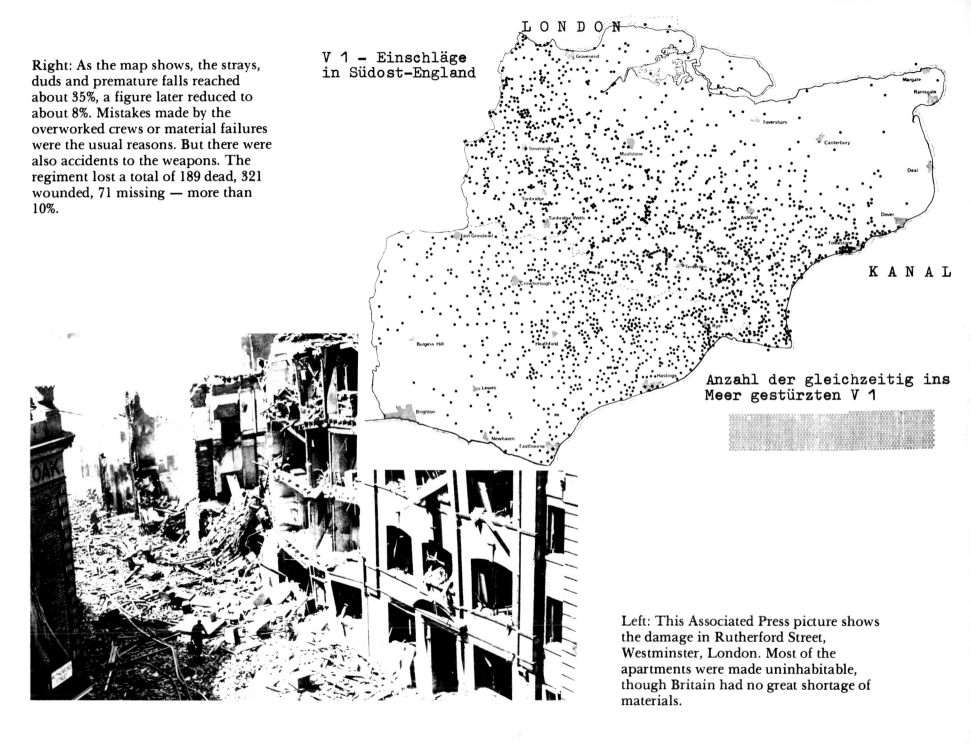

V 1 – Einschläge in Südost-England

L O N D O N

Gravesend

Margate

Ramsgate

Faversham

Canterbury

Sevenoaks

Maidstone

Deal

Tonbridge

Tunbridge Wells

Ashford

Dover

East Grinstead

Folkestone

K A N A L

Tenterden

Rye

Crowborough

Burgess Hill

Heathfield

Anzahl der gleichzeitig ins Meer gestürzten V 1

Hastings

Lewes

Brighton

Newhaven

Eastbourne

Left: This Associated Press picture shows the damage in Rutherford Street, Westminster, London. Most of the apartments were made uninhabitable, though Britain had no great shortage of materials.

Production

Left page: This view of the big assembly hall of the VW works at Fallersleben gives an impression of the series production of cells. Despite production of 23,748 cells in 1944, production was not enough for regular use.

Right: Since 83% of all cells came from Fallersleben, the factory was bombed again on August 5, 1944. Through decentralizing, its share was reduced to 35.4% in 1945. But production in the "central works" was high.

The supply of rockets to the front was usually done by rail, despite constant air raids. Yet the regiment, with two units of 8 to 18 launchers, fired 5790 rockets at Antwerp, Brussels and Liege between October 20 and December 31, 1944.

Right page: The V1 transporters of Flak Regiment (W) 255 had a bad time of it. With shorter rail connections, the regiment's three units of 35 to 47 launchers launched some 9700 missiles at Rotterdam, Belgium and London from January 1 to March 27, 1945. The 15,000th rocket was launched on January 12, 1945.

Underground Transfer

Above: Transfer of production to an underground facility in the southern Harz by the "Economic Research Company" (Wifo) took six weeks, through the use of all available means, including KZ prisoners and SS full powers.

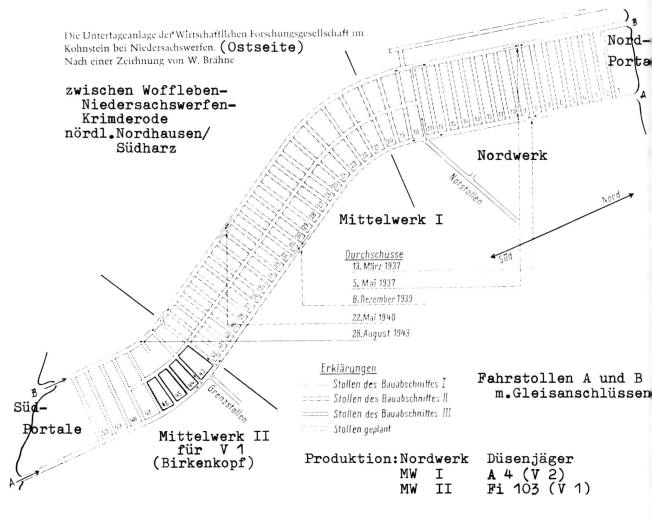

Die Untertageanlage der Wirtschaftlichen Forschungsgesellschaft im Kohnstein bei Niedersachswerfen. (Ostseite)
Nach einer Zeichnung von W. Brähne

zwischen Woffleben-
Niedersachswerfen-
Krimderode
nördl.Nordhausen/
Südharz

Above: Of the completed galleries in the Kohnstein, V1 production was given the four galleries 43 to 46 not far from the south gates, directly as rail connection A, near the production sites of the V2 and jet fighters, the most modern weapons.

Right page: Here the cramped nature of the damp and cold galleries for the assembly of the long and heavy rockets is made clear, quite apart from the unfavorable ventilation and lack of daylight. As of November 1, 1944 the "Middle Works", under contract to the VW Works, combined 16 branches and suppliers for the complete final assembly of the V1. The RKL price was RM 5000 apiece.

Cruising Rockets

Right: American copies took off from rocket sleds, seen here at Holloman AFB near Alamogordo, New Mexico, not far from the nuclear test center.

Below: The US Air Force launched so-called "JB-2" (for "Jet Bomb") from B-29 planes, as here from Eglin Field, Florida. The takeover and further development into the "Cruise Missile" of today in thirty years was obvious and not difficult.

Technical Data

	Combat Model 1944	Long-Range Model 1945	Planned
Overall length	7.742 m	8.874 m	
Fuselage length	7.16 m		
Fuselage width	0.823 m		
Motor length	3.66 m		
Max. motor width	0.571 m		
Speed	645 kph near ground	765 kph	800 kph
	575 kph at 3000 m	improved power plant	Feb. 1945
Range	257-286 km	370 km	500-650 km
			not used
Payload	847.11 kg	453.49 kg	
Wingspan	4.90 m	wood wings 5.7 m	
		metal wings 5.39 m	
Empennage span	2.00 m		
Take-off weight	2,200, 675 kg empty, 550kg fuel		
Propulsion	Heinkel catapult (steam-powered) with Walther fuel instead of Rheinmetall starting sled with powder rocket		
Flight time	30 minutes		
Powerplant	Argus 109-014	Argus 109-044	
Controls	Askania autopilot with 3 gyrocompasses over 2 axes, barometer by pneumatic elevators & rudder		
Cost apiece	4-5000 RM (1/10 of A4/V2)		
Production time	280 to 300 work hours		
Quantities	1944: 23,748	1945: 6509	
Designations	German:		
Fieseler Works	Fi 103 i		
Reich Air Ministry	FZG 76		
Volkswagen Works	Cell		
Prop. Ministry	V1		
Camouflage names	Cherrypit, Crow, Junebug, Richard Device, D Train		
Allied:	Flying Bomb, Robot Bomb, JB-2, Buzz-bomb, Thunderbug, Doodlebug		
Opponents	Tempest, Typhoon, Gloster E28/39, Gloster Meteor		
Successor models	Soviet Union	J Series medium-range rocket copies	
	Britain		
	France	Lacrosse, Malaface, Chasseur	
	Sweden	Robot 315	
	USA	Martin B-61 Matador, Regulus, Regulus I & II, Mace, Hound Dog, Snark, Cruise Missile	

Minor deviations in the data result from secrecy, type variations and conversion to American measurements.

Today's Relatives of the V1 Cruise Missiles

They are unmanned cruising aircraft. They fly near the ground, nearly at the speed of sound, for up to 2500 kilometers. They correct their course automatically according to preprogrammed landmarks. An electronic targeting system gives them great accuracy.

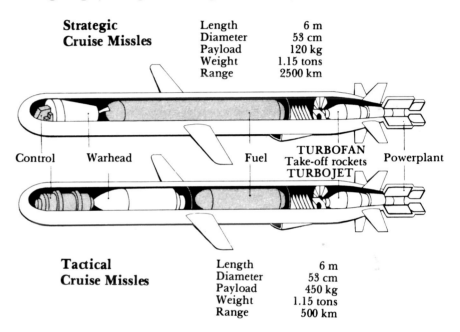

Strategic Cruise Missles		
Length	6 m	
Diameter	53 cm	
Payload	120 kg	
Weight	1.15 tons	
Range	2500 km	

Control Warhead Fuel TURBOFAN Take-off rockets Powerplant TURBOJET

Tactical Cruise Missles		
Length	6 m	
Diameter	53 cm	
Payload	450 kg	
Weight	1.15 tons	
Range	500 km	

The Cruise Missiles exist in three forms: Land-based (GLCM), sea-based (SLCM) and air-based (ALCM). They have only one warhead. Multiple warheads were expressly avoided. On the basis of their dimensions they are easy to transport, have many uses and are also suitable for tactical tasks. They are launched outside the range of enemy weapon effect. They can circumvent zones of concentrated air defense with correct programming of their flight path. Cruise missiles can thus take over the tasks of aircraft where the use of aircraft is risky because of strong air defense, but they cannot fully replace aircraft.

Since the mid-1960s, the Warsaw Pact also possesses cruising missiles. Almost all its larger warships are equipped with them.